Charlie + Nancy Jackson

YOUR SOLUTION TO FARM FINANCE

MARY JO IRMEN
FARM FINANCE CONSULTANT

FiscalBridge Publishing

Farming Without the Bank

ISBN: 978-0-9907052-0-8 (paperback)
ISBN: 978-0-9907052-1-5 (ebook / ePub)
ISBN: 978-0-9907052-2-2 (ebook / Kindle & Mobi)

Get the right digital edition for your favorite eReader: Get the ePub for iPads and B&N Nook, and Mobi for Kindle and the Sony eReader.

FiscalBridge, LLC—Farming Without The Bank
2000 Schafer St. Suite E
Bismarck, ND 58501
maryjo@fiscalbridge.com
www.farmingwithoutthebank.com
701-751-3917

Cover photo by iStockphoto © Jir
Book design by DesignForBooks.com

Printed in the United States of America.

I have dedicated this book to my parents,
brother and sister-in-law.

Without my parents I would not have had the opportunity to grow up knowing where my food comes from, the work it takes to get it there and the wonderful life you live on the farm despite the gamble of it all. Even though I disliked chores at the time, today they are looked at as a blessing.

Without my brother and sister-in-law our family farm may not have continued through the next generation. Their hard work and dedication must be recognized and appreciated by those of us who are not out there doing what they do every day to keep the farm going.

Contents

Acknowledgments

Though it shows my name as the author of this book it was not written without the guidance of many great people in my life.

First and foremost without R. Nelson Nash and his book, *Becoming Your Own Banker,* I would not be where I am today. This book and Nelson's encouragement have changed the direction of my life. All of those questions I had for so many years about if there were other options outside the traditional 401K or IRA's were answered.

Wade Borth has to be given equal credit. From the very beginning he was one of the few who shared my passion for farmers and ranchers. Wade understands the importance of sharing and teaching our farmers this process. He has been there to answer what may seem like a thousand questions. He has given me countless hours of his time with his insight and content proofing of this book. Truly his name should be on the front cover right beside mine. I could not ask for a better "farm kid" colleague to share this adventure with.

If you were to ask anyone from my high school days if I would ever write, they would laugh. Putting thoughts on paper is not hard for me. Putting thoughts on paper and writing so you can make sense of it all is not easy. A huge thank you to those who helped me proof and edit this book. It was not a small task for them, as I tend to write like I talk.

I cannot leave out all those who have given me their time and efforts to share in this book, either through proofing it or answering publishing questions. It truly shows I am not alone, and we are all here to work together and help each other out.

Finally the biggest thank you must go to my amazing husband and family, who are always there to encourage and just plain put up with me and my ideas. There is no way I could do this without their support.

Agriculture Progression

Introduction

Growing up on a farm we had purebred cattle and did some farming. Cattle were the staple to our operation, and I remember sitting at the supper table listening to my dad talk about how many pounds a day the bulls needed to gain to be ready for sale day. He would sit with a calculator and punch numbers as he ate lunch, figuring out how much to feed and what kind of feed. Each and every day, consideration was giving to how to maximize profits, including plans for making sure each cow was bred, minimizing the risk of losing calves, and ensuring proper daily weight gain until market. If any of these factors went wrong, there was not going to be much left over after the bank note was paid. Everything factored into making that note and how much dad had AFTER Mr. Banker was paid his share.

Today my parents no longer have cattle, but they continue to farm with my brother and sister-in-law and, like every other farmer, it all comes down to the profit they can make AFTER operating costs, including loan costs, are paid.

In a few short decades agriculture has progressed from horses and plows to top notch equipment with GPS systems. This progress is not going to stop, and farmers can't afford for it to stop. Farmers are feeding more people today than ever before because of the advancements in technology of equipment and seed production. Efficiency in production is addressed daily now. **It is time to address efficiency of farm financing.** What we are not seeing is advancement in the way these farming operations are financed. Farmers continue financing their operations through banks, just as they did 100 years ago.

Growing up on a farm and ranch provided me the ability to understand on a deeper level how agriculture is financed and why it needs to change. Loan officers are typically like family members because without family, in most cases, these farms would not be operating. You, as a farmer, are forced to look past the amount of interest and control given to the banks because you aren't given any other options. Instead, the borrowing and worrying cycle continues from year to year.

The Future of Farm Finance

This traditional system of financing a farming operation is not the only way. You are not alone in this debacle; every person that borrows money is in this position. It doesn't matter if you are buying a car or buying a combine. **If you are borrowing money the bank is making money, off the interest. Banks feed you with affordable payments so you forget the interest factor.** The difference between you and the guy in town buying a house is that you give banks much more interest money. Borrowing $300,000 over 30 years to buy a house in town is pocket change interest compared to you who maybe borrowing $750,000 every single year to operate your farm. As my dad told one of his siblings who doesn't farm, "I go through more money in a year than you will go through in your lifetime." That is a pretty accurate statement. You deal with large sums of money and you pay a high price to use that money.

The Infinite Banking Concept® described by R. Nelson Nash is a method that can change the way you finance your agricultural operations. It puts you in the position of being your own banker and puts what would have been lost money back into your pockets, as well as giving you back the control you lose to the bank.

The Infinite Banking Concept® is just that—a concept that applies a process where you have infinite strategies to break away from the traditional lending process.

I have been teaching this concept for several years, and when I began I immediately thought of how this could help the agriculture industry. I started with a couple of clients in the agriculture industry and saw how it was helping them eliminate the bank from their equations in just a few short years. In the beginning these particular clients were employed off the farm in addition to farming. In those few short years I have seen some quit their outside jobs to solely farm and others increase their cattle herd. Both of these things were not happening before they started using this concept. This has provided them with the freedom and flexibility they needed to move forward.

Staying Open-Minded

While you continue reading think of this: "You can lead a human to knowledge but you can't make him think." While reading this there is one requirement: you must use your reason and logic to envision how applying this concept can change your own situation. Maybe it will help you turn over the operation to a son or daughter more quickly, or maybe it will help you grow your assets more quickly. If you already think you know it all and are not open to hearing something new then you may as well close the cover now and pass it on to someone who is willing to learn. This book is designed to help you challenge your own assumptions and think about how you can break free from some of the costs for doing business to which you have likely become accustomed.

No one needs to hear more guru financial advice. What you do need is a new strategy of proven success to finance all those things you purchase. This book will provide solid information to help you think differently from what generations before you were taught regarding how to finance their agricultural operations.

You will find in this book the information you need to advance your farm financially. Just as you would shake your head at someone farming with horses and plows, you may shake your head in disbelief that you didn't hear of this sooner. I hear "Where were you 20 years ago?" often in my industry. Depending on your age, you may find yourself saying the same thing.

What you will not find in this book is fluff and sugar-coating. I know you have a lot to do, so I get right to the point. You will see how it works, why it works and what makes it better than what you are doing now.

Your Choices

If you are still reading then I will presume you are open-minded. After you have completed the book you can choose to do what you want with this information. You will fall into one of these five categories:

1. You will start the process with a certified practitioner and apply the concept. You allow a certified practitioner to teach you the process and use them as your strategy consultant. Result: SUCCESS.
2. You will start the process with a certified practitioner but won't apply the concept to your operation. Result: You have taken a baby step forward but won't be going anywhere fast. As

Nelson Nash says, "you use it or lose it." If you don't use the concept you forget why you even started.

3. You go see your friend/family member that sells traditional insurance and get started. Result: I wish you luck, as 99% of those in the financial industry have no idea how to correctly setup and implement this process.

4. You read it, like it and never start. Result: Nothing. Charles Dickens said, "Procrastination is the thief of time, collar him." In this case procrastination is the thief of money.

5. You don't believe it works because your relative, who never heard of it or studied it, said so and you take their word for it. Result: Continued lost interest money and control to banks and you keep the same conditioned thinking banks have taught you over the years.

Just like in farming, if you don't plant the seed nothing will grow, and if you plant the seed and don't tend it, weeds will choke out the growth. This process is the same way. You must first plant the seed and then tend it. Only then will you see the results of your hard work.

Let's start planting some seeds.

Conditioned Thinking

Change Your Thought Process

To be honest I was not even going to add this to the book. However, after visiting with a number of farmers, it was clear those looking to change what is being done are the same ones who are already thinking differently. I also noticed some farmers are not bothered by certain things such as paying interest or having their "paycheck" come with the bank's name on it. (For lack of a better term I am going to call the sales of your commodities your paycheck.)

Not being bothered by paying interest and having Mr. Banker's name on your check is simply conditioned thinking—conditioned to think there is only one way to do something because it's always been done this way. As Albert Einstein said, "The important thing is not to stop questioning. Curiosity has its own reason for existing." Never questioning the financial process doesn't allow a lot of room for improvement. It would be like never questioning farm practices.

We have been magnificently conditioned to think we need banks. Just like trainers who condition elephants to believe they are inescapably tethered by a small rope, many of us believe that we are tethered to banks as the only way to manage and grow our money. We learn from our parents from a very young age to borrow from banks . . . and if nothing changes, our children will learn the same from us. You are not tethered. Learn how to be financially free from banks and the uncertainty of the stock markets.

In order to move forward you must change the way you think. If you have your farmer cap on then I need you to also put on your business cap. Farming is a business, and all businesses are operational for the sole purpose of making money. You as a farmer are most likely making money; however, your bottom line is always affected by expenses. If you reduce expenses, your bottom line will be better. Limiting the flow of your money to others in the form of interest is just the beginning. You can become your own bank by beginning to think like a banker.

On page 34 of the book *Becoming Your Own Banker*, R. Nelson Nash, talks about the person who thinks they know it all and that there is nothing else to learn. Nelson calls this the Arrival Syndrome. Daniel Boorstin said it well, "The greatest obstacle to discovering the shape of the earth, the continents, and the oceans was not ignorance—it was the illusion of knowledge."[1] You may hear me talk about the Arrival Syndrome throughout this book. **If you believe you have arrived, your mind is closed to anything new.** I will assume that is not the case since you are still reading.

Conditioned thinking combined with the Arrival Syndrome can be very dangerous. If you truly arrived in your thinking, you would still be farming as you did 30 years ago and I am confident you have progressed in your farming. The funny thing is there will be plenty of people who have arrived without ever even giving the concepts presented in this book a second thought. We all know people who may dismiss it and figure they don't need it because they are making a profit. Never mind the fact that they could make a bigger profit, feed more people or just plain old learn something.

Thomas Edison said, "Five percent of the people think; ten percent of the people think they think; and the other eight-five percent would rather die than think." When we look at this do those 85% not think, are they conditioned to not ask questions or have they arrived and believe they know everything they need to know? There is a huge difference between not questioning and knowing it all.

---

1. Daniel Boorstin Nash, N. R. (2009) *Becoming Your Own Banker*

Living Within Your Means

Another thing Nelson Nash talks about in his book on page 28 is Parkinson's Law. C. Northcote Parkinson said, "A luxury, once enjoyed, becomes a necessity." An example of this would be air conditioners in your tractors. Farming without air conditioners, once the standard, would now be a horrible experience, right? Once you have a tractor with air conditioning you would never think to buy one without. What was once a luxury is today a necessity. The list could go on and on to include all kinds of designs and technologies ranging from heated leather seats to covered cabs to GPS systems.

Not every luxury is a necessity. Heated leather seats are not a necessity. You may be more comfortable during your ride, but it doesn't make the vehicle more efficient. GPS systems will soon move from a luxury to a necessity. All these luxuries come at a price. Those who are not willing to get a grasp on Parkinson's Law will see their finances struggling just to keep up with all the luxuries. Those who get ahead in life are those who control Parkinson's Law. Money spent on luxuries you can't afford will keep you controlled by the banks.

You all have the neighbor who has to have the brand new tractor yet s/he is one bad year away from bankruptcy. You may also look at Parkinson's Law as keeping up with the Joneses. From just these two things you can see the way you think affects every part of your life from your attitude to your pocket book.

Use and Storage of Money

Two factors you are not conditioned to think about are honest use and correct storage of money.

Honesty and Self-Respect

By honest use, I mean stop stealing from yourself. I am willing to say you would never steal from someone because you know it's illegal and wrong. I am also willing to say **you are stealing from yourself every day and you don't realize it.**

If you are like the majority of people, you have been taught how to use money incorrectly. When you buy something with credit cards or borrowed money you simply pay that money back plus interest. In fact, you may not even ask what the interest rate is on that borrowed money. You need the money and can afford the payment, so there is no need to question or concern yourself with anything more. What you may not realize, as Nelson points out in his book, is that 34.5% of every dollar spent goes to debt service (interest and fees). This number is all relative to the size of your operation. It doesn't matter if you are a beginning farmer or one who has a three million dollar annual operation; you've lost interest to the bank.

On the other hand, when you use cash to purchase something, you don't even consider paying yourself back PLUS interest. Heck, you may not even consider paying yourself back because you don't think of it as a loan.

If you are a saver, you will continue to save by refilling your savings account, but you will most likely not pay yourself interest. Why wouldn't you give your money the same respect you give the credit card companies' or banks' money? Do you think more of them than you do yourself? If you are going to use your own saved money, put the borrowed money back PLUS interest. Treat every transaction as if you put it on the credit card or borrowed it from the bank. Imagine yourself as the banker charging interest for the use of that money. If you are willing to give someone else interest for the use of their money but not pay interest to yourself, you are, in essence, stealing from yourself.

To further my point, let's look at an example. You want a 4-wheeler and want to pay cash for it so you save $5,000. In order to have this money in savings you put it there through monthly savings contributions. Once the $5,000 is there, you buy your 4-wheeler and your account is now at zero. You decide not to continue saving because you met your end result. If you are thinking like a banker, you will be diligent and continue making those monthly contributions into your savings account to replace the $5,000 you took. If you think as much of yourself as you do the banker, you will charge yourself interest and replace the $5,000 PLUS interest.

When you pay yourself back the $5,000 plus interest your account grows by the interest amount. You've now created wealth with money you have just used to buy something else.

Storing Money

Now that you are thinking like an honest banker, we can move into storage of money. Did you have a savings account designated just for that 4-wheeler? If so, I want to know why that or any account is set up for just a single intention?

This compartmentalizing of money has been taught to us from the time we are little. Our parents did this. We were also shown how to do this in school, and even the financial gurus of today still continue to condition our thinking to compartmentalize money. We are told to keep separate accounts for each goal (Christmas fund, college fund, etc.). Now, I can understand this logic when you are separating business from personal expenses, but that is the only reason.

You do not need to compartmentalize money if you are honest with your money, like we just discussed. Money flows when you continue the savings cycle by taking and replenishing so you can take again for another use.

To illustrate this, think about the main water supply coming into your home. Let's say that it all comes from one big pool of water—the well. However, in your home you have many faucets between the bathrooms, kitchen, laundry room and outside. If you want a glass of water and you fill that glass in the kitchen, you just took water from the well the same as if you would have filled it in the bathroom or outside. Every time you fill a glass of water, the well gets refilled from its source so you have water again for next time.

Money works the same way! You have earned income that supplies money to your household on a regular basis. This money can be put into one big pool and used for many things IF you are honest and replenish it. Just as you would run out of water if it didn't replenish, you will run out of money to use. If you compartmentalize money coming into your home it would be like only using kitchen faucet water for drinking. If you wanted to wash dishes you would have to have a separate well for that.

Now let's tie honesty and storage together to see how it works. You have a general savings account and have been saving for next year's vacation and currently have $5,000 in there. You need $1,000 worth of new tires on your pickup, and you borrow money from your vacation fund account to pay for them. Over the next year you are honest and you put $1,000 back plus 10% interest into the savings account. A year from now you have $5,100 to go on vacation. See how that vacation fund was not just for vacation? That money was used for both vacation and necessities.

Money Working for You

Storage is also about making sure that money does something for you while you have it. You can store money in the bank or under your mattress, but it doesn't earn interest in either of those spots.

Your money should be treated like your land. Your land value goes up while you farm that land to make more money. You should be doing the same thing with your money by putting it into a system that allows you to earn something while you use it to make more money.

Being honest with yourself allows you to think more freely about money. So many of my clients come to me stressed out that they are broke and can't buy anything. Yet they have three different accounts, all designated for a separate purpose. Because they don't have an account for tires, they feel as if they can't afford the tires. In fact, they do have the money. They just don't have the correct mindset that they can repay the money back so they can use it for both.

This honesty mindset will also allow you to manage debt better. If you don't have the obligation to repay yourself, you tend to overspend. If you are honest and repay yourself you realize you can't buy it if you can't afford the payment back to yourself, just as you wouldn't buy something you can't afford to repay to a bank.

PAYMENTS AND INTEREST

Volume of Interest

Remember, the typical conditioned thinking is that if you can't pay cash for it, just borrow the money and it's okay as long as the payment is affordable. As I said earlier, your purchases are either made with cash or borrowed money and put on payments.

What you are not taught to think about is the **rate of interest** you pay is not as important as the **volume of interest** you pay. You go out and buy based on the lowest rate and the most affordable payment. It doesn't matter if you have a 2% interest rate on one big loan or ten small loans. The volume (amount) of interest is the same and is what adds up without you noticing.

If you bought a tractor for $300,000 at 6% interest over seven years, that tractor would have cost you 354,890.13 at the end of those seven years. You just lost

54,890.13 in interest to the bank . . . ON JUST THIS ONE PURCHASE. How many more purchases do you have out there like this one that you are paying interest on. Even a 2% loan is lost money.

You have to stop looking solely at the interest rate and payment and start adding up the volume of interest you are losing each year. Then **think about that lost interest over your lifetime.** In the farming industry you can't stop buying things until you stop farming, but you can stop giving interest money to someone else and instead pay it back to an entity you own and control. With the amount you spend on operating, land, equipment, cattle and feed, anything saved is a further advancement for your operation.

Some of you are saying interest charges are not a big deal because you get to write it off. Even though I understand the thought process here, I am going to ask you to consider the following: If you are in a 25% tax bracket and pay 15% Social security that means your write-off is only 40% of the taxes you paid.

Interest on tractor	$54,890.13
Interest saved through taxes	$21,956.05
Interest lost after taxes	$32,934.08

That is still a loss of $32,934.08. Not to mention the $300,000 in principal you lost access to. You know as well as I the interest lost is all in relation to the size of your operation. Regardless if it's $5,000 or $250,000 a year, it negatively affects your cash flow.

I can't say it enough—it's not the percentage of interest, it's the volume of interest. If you are an established farmer, I challenge you to add up the volume of interest you have lost over the years. If you are a new farmer, I encourage you to estimate what you will lose over your lifetime. What would this money have done or do for your operation?

Here is a story of arrival syndrome at its fullest. I was visiting with a farmer and we were talking about the interest he pays to the banks. I asked him, "Doesn't it bother you that you pay $60,000 a year in interest to the bank?" His reply was, "No, if I can pay $60,000 to make $300,000 why wouldn't I?" My response was, "Because you just lost an additional $60,000 you could use for your operation. You could have $360,000." This gentleman had arrived. He did not want to hear anything new; he was comfortable with the way he had been financing the farm for the last 45 years and loved his banker. Had he stopped long enough to think about it, he would see that Mr. Banker will get $600,000 in the next ten years. No one but him will ever know what he has already paid in interest. This man is obviously a great farmer or he wouldn't be making this kind of money. What has happened is he has been very well conditioned to think borrowing from banks is the best way to go.

As George S. Patton said, "If everyone is thinking alike, then somebody isn't thinking." Start thinking of your payments as two parts and keep track of how much volume in interest is leaving your hands.

Cash Purchases

Lost Opportunity Cost

You have likely been conditioned to think of cash as king . . . to save your pennies in a savings account until you have the cash you need to make your purchases. This idea is not wrong; it is better than losing interest through financing as we just talked about, but it is not King. Instead, I like to say cash is Queen.

You may be using cash to operate, but you are still financing the operation. I am going to quote Mr. Nash because he says it perfectly in *Becoming Your Own Banker*, "You finance everything that you buy—you either pay interest to someone else or you give up interest you could have earned otherwise." This in finance technical terms is called lost opportunity cost. It can be compared to cash being like hired help. When you spend cash and don't replace it, you just fired your hired help.

You know as well as I do that you are always looking for the most efficient way to farm, to make more profit on each acre. Not using the full potential of a dollar is the same thing. You paid cash for operating and in turn gave up the ability to ever use that money again or have that money earn anything.

Compound Interest

The value of the dollar does not appreciate as land does, but you can put that same dollar somewhere that allows you to earn interest on it and slow down the devaluation. Not just any old kind of interest but compounding interest.

You most likely have been taught not to rack up a big bill on your credit card because the compound interest charges get out of control. This is true. Those companies are making millions off of consumers paying compound interest. What you want to do is be like the credit card company and make compounding interest. Albert Einstein said, "Compound interest is the eighth wonder of the world. He, who understands it, earns it . . . he who doesn't pays it." This statement speaks volumes when you learn what compound interest can do when you make it work for you.

Compounding money is much like creating a cattle herd. If you sell a cow and don't replace her, you have lost out on all her future offspring over the next fifteen years. You know the importance of multiplying your herd.

You may have been exposed to compounding in the terms of twos. Two turns into four, four into eight, eight to sixteen, sixteen to thirty-two, thirty-two to sixty-four and so on. You can see as the numbers get exponentially bigger with the doubling.

Below is an example of what compounding interest looks like on a onetime $100,000 earning 4% over 50 years. You would have a total of $710,668.33. Had you used this $100,000 to purchase grain bins and paid cash for them you would have lost the opportunity to have the extra $610,668.33. Instead you just got the bins.

Current Principal	$100,000.00
Annual Addition	$0
Years to Grow	50
Interest Rate	4%
Future Value	$710,668.33

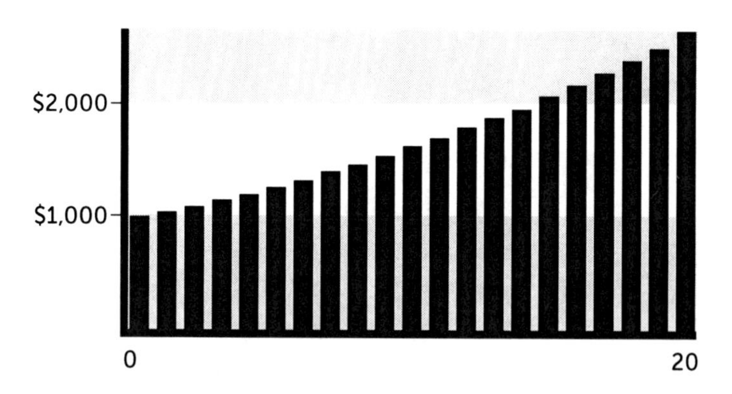

Amortizing Interest

Not all interest is created equal. There is also amortizing interest, and this is the type with which you are probably most familiar. Loans for land, operating, equipment and vehicles are amortizing. You have a set amount of interest charged onto the loan, your payment includes mostly interest at the beginning of the loan and you pay the bulk of the principal closer to the end of the loan as you make your payments. This way you pay down the interest over time.

As you can see in the following charts, had you financed the bins the amortizing interest lost was far less than what you would have made compounding that money.

Loan Amount	$100,000
Loan Term in Years	7
Interest Rate	4%
Total Interest Paid	$14,817.97

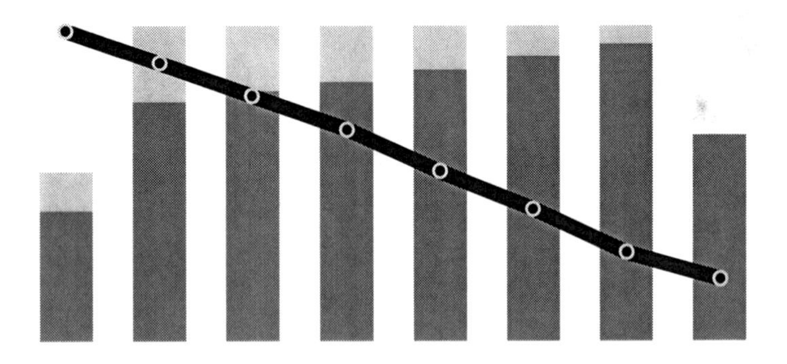

These two types of interest are what is important when we begin discussing why this concept works so well for you as a farmer.

Cost of Using Cash

The graph below illustrates the power of compound interest versus the use of cash. If you were to save $15,000 each year and after four years take that money out of savings to buy a new pickup then repeat the cycle for 30 years, you would have LOST $760,029.24 of compounding interest.

Each time you take money from savings you stop earning interest, and each time you must start over just to earn a little at a time. Using a tool which allows this process to continue even while you purchase things is the key to success.

Annual Contribution	$15,000.00
Rate of Return	4%
Time (Years)	30
Number of Purchases	6
Years to Save for Purchase	4
Results	
Future Value	$841,274.08
Lost Interest	($760,029.24)

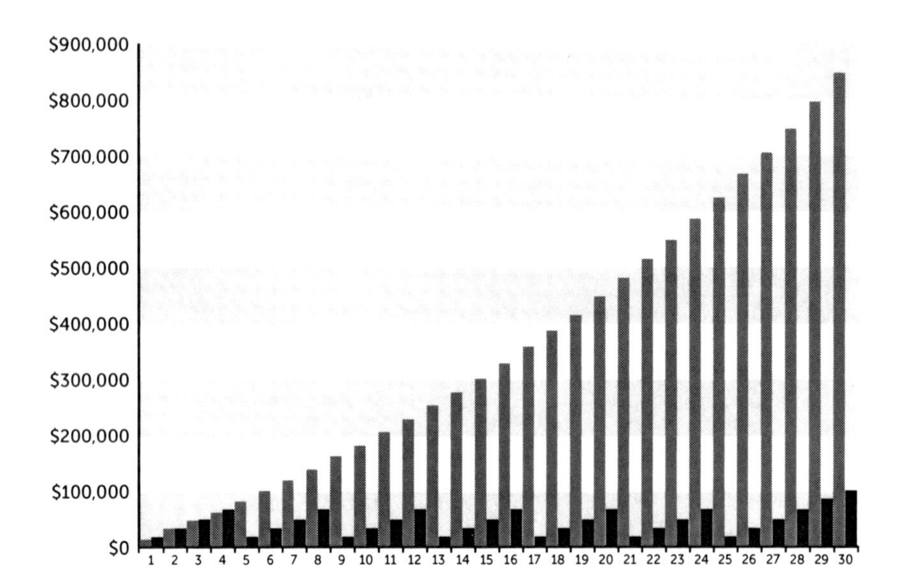

Since cash purchases and financing are common, what is probably a new concept to you is the idea that you can keep your money compounding while you use it, again, just as you do with your land and cattle.

Keep this chapter handy because you may need to reference it later for understanding how compounding interest increases while amortizing decreases.

Banks Control

Financial Progression

Farming and ranching have come a long way, but as stated earlier, the way you finance your operation has likely not changed. With today's operating expenses, finance has become the most profitable part of your operation, and you are most likely letting someone else be in charge of your finances. Each and every year you depend on your banker for money to operate as well as buy land, equipment, cattle and feed.

Bankers almost become a part of the family. In most cases the better the relationship and more trust you have with bankers, the easier it makes the borrowing process. I know farmers who believe paying interest to banks is okay because it's the way they make a living and it's just been done this way forever.

At what point does it start to bother you? I have had a couple farmers come to see me because they had reached their limit of frustration and were tired of losing over $150,000 in interest to the bank each year. Their frustration kicked in when this interest money was the difference between making a profit or having a loss for the year. I am sure at some point you have been in the same position. It's easier to stomach the lost interest when you're making a profit and getting along with the banker.

Accessibility

Borrowing money is not equally accessible to all farmers. It truly depends on from whom you are borrowing money and what you look like on paper. Some banks will require every receipt to validate the money was spent in the best manner possible. This is understandable because the bank wants to be sure if things go badly they will have enough equity to recover their losses. However, what is interesting is a fair amount of bankers are sitting behind a desk and have no idea how you operate, yet they are telling you what to plant, how many cows to buy, and when and what to sell. When you borrow money from banks you have essentially agreed to a business partnership where you no longer independently control the farm. The bank is your partner who has majority power, in most cases, because they put up the funds.

I visited with a young lady who was getting started in the ranching business. Her banker approved a loan of $20,000 for her to buy ten cows. She asked if it would be okay to buy twelve cows with that money rather than just ten, as she was able to work out a deal with the seller. The banker told her no, she was approved to purchase ten cows only with the $20,000. This makes no sense. If she was approved to borrow $20,000 and able to buy more cows, that just means a bigger herd and more profit. This is a case of a banker controlling the entire operation from behind a desk. This young lady has little control of what is supposed to be her operation. She is not working for herself; she is working for a bank that has little risk because she has the cattle as collateral.

Over the years I have also visited with established farmers who are able to walk into the bank and request money without question. As an established farmer you may be more comfortable with this borrowing process because you feel you have the upper hand and the process is very familiar to you. To some extent you do have a little power because you have established equity and years of knowledge behind you. Even this little bit of power may change if the bank changes ownership. Regardless of your position, just like the young lady above, you still give up some control every time you borrow for operating costs.

Your Livelihood is Collateral

Every time you get a loan, the bank has control and has rights to your farm as collateral. You, on the other hand, just put up your livelihood, leaving you with all the risk. Should you default on the loan, the bank takes your assets leaving you with no way to farm.

I had a farmer tell me, "Every farmer is two to three bad years away from bankruptcy." Beginner or established, should you have a couple of bad years in a row, you are no longer the one with the upper hand. The potential is still there to lose everything because the bank has the legal right to start selling off the equity to cover the payments you are unable to make.

This is just what happened in the 80's when interest rates were high and farmers were buying land. It could not be sustained and bankruptcy became rampant. In order to get out of the mess, farmers had to sell off their land.

Since visiting with the farmer who made the statement about bankruptcy I have had many other farmers, small and large, confirm this to be true, which only proves the point that how much you borrow is relative to the size of your operation. Regardless if you have 15,000 acres or 1,200 acres, 50 cows or 500 cows, you are all in the same boats. Each boat is just a different size. The little boats may have little holes and the big boats have big holes. but they will sink equally as fast.

Even as I write this book I am seeing a flash back of the 1980's with farmers who come to see me. Commodity prices are dropping again and banks are pulling back quickly. Farmers who have made payments for years are having trouble getting loans for next year's operating and are being told to sell land. It is becoming clear very quickly these banks are not on the sides of the farmers.

Sharing Your Paycheck

It's not just about loss of control in decision making and your livelihood; there is also loss of control of your "paycheck." Even if you have a good year, you haul grain to the elevator and cattle to the market, but when you get your "paycheck" it has your name and the bank's name on it. You are not allowed to have this money you worked so hard for without the bank's permission. You must go from the elevator and/or sales barn to the bank just to have Mr. Banker take it as a payment or sign off on it. Banks get their money first before you can have yours. To top it off, if your loan is not paid off completely, some banks ask you to disclose what you plan to do with the money they release to you.

When I get my "paychecks" I am not required to have my mortgage lender's name on the check, nor am I required to pay them first. Should I have more money than my mortgage payment, I am not required to tell them where I am spending the excess cash. Why is money lent to farmers treated differently than money lent to non-farmers?

It all boils down to control. When taking money from a bank to operate, you are giving up control in some capacity. What if high interest rates, like those in the 1980s, return? Will you be in a position to continue farming or be stuck owing the bank money? It is your operation and the control needs to shift from the bank to you. To take back full control you need to become the farmer and the bank. I am guessing it will be a lot easier to deal with yourself when and if you have to.

Now you are aware of the conditioned thinking you may have as well as the control over your assets the banks have. In order to change both of these things you will need to understand how the Infinite Banking Concept® can work for you. The rest of this book will cover the benefits and strategies of this concept and how the process works.

Infinite Banking Concept®

What It Is

The Infinite Banking Concept® is a concept that uses a simple process where you are able to keep your money working while you use it. As with anything, this concept requires a tool, and the tool used for this is a dividend paying whole life insurance policy. A dividend paying whole life insurance policy is the perfect tool for creating a personal bank and providing death benefit for your family. There are other tools that can be used, but long term whole life insurance is the only one that allows you to have the most efficient financial system.

A strong, solid foundation has been built around whole life insurance. It has been around for over 150 years, succeeded and done well. That is important to recognize when you start comparing it to other things like the Federal Reserve which has been around for less time. Only about

100 years ago, the Federal Reserve was formed and took control of dollar valuation. The current value of our dollar shows how well that has done . . . a stock market that can't find stable footing.

The guarantees provided by properly structured dividend paying whole life insurance from a mutual company offer the perfect banking tool. Your premium will not change throughout life. Your death benefit will increase to provide for your family. And your cash value will be available for you when you need it.

Term Insurance

You, most likely, know more about term insurance because it is a fairly inexpensive option for death benefit. Term insurance fills only one need, which is death benefit, the most inefficient tool in your box. Remember we are looking for efficiency. The misconception is that term insurance is "cheap," and that is true until you hit age 66 when premiums increase. What you do not hear is at age 66, coverage that was once around $1,000/year is now closer to $16,000/year, if in good health. The main reason you do not hear this is because you are told you should not need it because you will have no debt after age 65. This is not even close to a true statement for farmers, who are often still farming and carrying debt at age 65.

If you are concerned about the cost of whole life insurance, compare the long term cost of term insurance through age 80 to the cost of whole life. Your premium totals are actually LESS with whole life. This is because term insurance has no guarantee your premium will stay the same after your term is up, and, as you get older, the premium sky rockets because you are closer to death. Early in a term policy the company knows statistically you will most likely not die, and they will not have to pay. This is not true later in life. It seems silly to actually say it, but the older you get, statistically the more likely you are to die, and this is factored into the actuaries' calculations.

Whole Life Background

Most people under the age of 60 have not heard of whole life insurance, as it was squelched with the boom of the stock market years ago. Dividend paying whole life insurance is a life insurance policy that will provide death benefit, dividends (non-guaranteed) and cash value throughout your entire life. A whole life insurance policy is also managed for the benefit of the policy holder, where profits are shared with the owners, which is you. These three parts are important to understanding how to make this concept work.

A whole life insurance policy is a contract much like your contracts for grain or cattle. When you decide on a premium amount, that amount stays the same throughout your lifetime.

Don't be confused with other permanent products such as Universal, Variable, Flexible Universal, Variable Universal and Equity Indexed, which also have cash value.

They are not the right tool for efficiency either. These types of policies are run for the profits to go to the insurance company (rather than the policy owner). Just because someone is talking about cash value doesn't mean they have the right tool. These other policies put your cash value and death benefit in a position where it can be affected by the markets' fluctuations. At the end of the day, if you are using this concept for financing, you want to have a guaranteed product that has little to no risk. Whole life is the best option you have that meets that criterion.

Example of Other Permanent Products

Below is an image of a client's Variable Universal Life policy. I have included this (with permission) to show you what happens with charges/fees inside these type of policies. As you can see, his premium per year is $11,300.04 for $500,000 of death benefit. Then you take a look at the Accumulation Value Summary section, which shows there are charges for Cost of Insurance, Premium Charges, Monthly Policy Charges, Policy Issues Charges, M&E (Mortality) Risk Charges and Additional Agreements Charges. If you add all these charges up, he is paying $6,803.72/year in charges. That means 60% of his premium is going to charges.

None of these charges are set charges; they fluctuate every year. The Cost of Insurance Charge is based on the age of the insured. As the insured gets older this charge gets higher, because the insured is more likely to die and the company anticipates having to pay. The company is going to make sure they get their share to pay out the death benefit, by charging more each year. It's like having a term policy that renews every year at your new age.

It gets worse. Look over at the next column where you see Accumulation Value as of 04/27/2014, Surrender Charge and Surrender Value. What this is saying is he has a total cash value of $28,960.39, but if he decides to cancel the policy he will only get the $13,035.67. The company is going to keep the rest as Surrender Charge. Even if he wanted to borrow against this money, he still only has access to the $13,035.67.

Policy Information

Owner:	**Policy Date:** April 27, 2009
Insured:	**Gender:** Female
Issue Age: 64	**Risk Class:** Non-Tobacco Plus Non-Tobacco

Summary of Current Coverage

Face Amount:	$500,000.00	**Annual Planned Premium:**	$11,300.04
Current Death Benefit:	$500,000.00	**Planned Premium Frequency:**	Monthly
Additional Agreements: Death Benefit Guarantee Agreement, Overloan Protection Agreement		**Billing Method:**	EFT
		Death Benefit Option:	Level
		Death Benefit Qualification Test: Guideline Premium Test	

Accumulation Value Summary

Accumulation Value on 04/27/2013:	$23,692.07	**Accumulation Value on 04/27/2014:**	**$28,960.39**
Premiums Received:	$11,300.04	**Surrender Charge**	**$-15,924.72**
Net Investment Gain/Loss**	$772.00	**Surrender Value**	**$13,035.67**
Premium Charges	**$-452.04**		
Cost of Insurance Charges	**$-3,868.36**		
Total Monthly Policy Charges	**$-96.00**		
Policy Issue Charges	**$-2,274.96**		
M & E Risk Charges	**$-93.45**		
Additional Agreements Charges	**$-18.91**		
Accumulation Value on 04/27/2014:	$28,960.39		

**May include unrealized loan interest credit

The next logical question is: Why are these types of policies even sold? My thoughts on the matter are these: The financial industry as a whole focuses on rates of return so you can make money fast—the get rich quick mentality that is really just a get rich quick scheme. Agents who sell these products are typically financial planners as well, and they are all taught rates of return in a whole life policy are not as good. Insurance companies push these other products based on that knowledge as well. No one is truly sitting back and running the numbers; they are only doing what they are conditioned to believe.

If you look at the rate of return in this policy, there was a -17.24% rate of return. And this was in a year when the market was supposed to be doing well. Yes, there was an increase in accumulation, but there was also a contribution of $11,300.04. Had he not contributed that, the loss would have been even greater.

Dividend paying whole life insurance is the only option that meets the criteria for this concept. In a whole life insurance policy you do not have these charges nor do you have a surrender charge should you decide to take or use the money. You need to build a banking system on a product that will give you the guarantees and have the stability of 150 years behind them.

Cash Surrender Value

As Mr. Nash says, "Your need for financing in life is greater than your need for death benefit." Who better to finance your purchases than you?

What is Cash Surrender Value

Even if you are familiar with cash surrender value (which I will refer to going forward as cash value) this chapter is one of the most important parts of this book, because most don't know how it truly works for self-benefit. Cash surrender value is the cash value in the policy. You pay a premium, then these premium dollars are accessible through this cash value portion of the policy to which you have complete access and control.

Cash value is there for these three reasons:

1 If you should decide to cancel (terminate) the

policy, the company will send you the amount you have in cash surrender value.

2. The amount you can borrow against should you want to take a loan.
3. The amount that can also be withdrawn from the policy while the policy remains in effect.

Access to Cash Value

What you will soon understand are the reasons you should borrow against this money and use it for financing everything you buy.

When you pay premium and have access to those premium dollars later on, it makes the premium feel like a deposit rather than an expense. This is why you will hear people refer to cash value as a "living benefit." You are able to use it now while you are alive and still provide a death benefit later for your heirs.

Cash value is **your** money. You have total access to it for whatever you want, whenever you want it, just like your savings account at the local bank. It is easy to get to. You call your agent, sign a service request form, send it to the life insurance company and the company sends you their money.

You then decide when and how you will pay it back. It's your loan and it's your decision. If you want to pay it back in a year's time you can do that, or pay it back in ten years' time. The decision is yours because you are in charge. The only thing you need to remember is it's your account, and you should be an honest banker as we visited about earlier in the book. Paying it back PLUS interest allows you more funds to borrow against the next time you need them.

Borrowing Against Cash Value

It may sound like a simple process of borrowing and paying back, but you must understand what is going on with your money while you borrow against it. Notice how I have said borrowing against the cash value. That is for a reason.

When you borrow against your cash value, the life insurance company leaves your money in your account and uses it as collateral while they lend you their money. **While you are using their money, they continue to pay you compound interest on your money because it was never removed from your account, just collateralized.** Remember how powerful compound interest was in that earlier $100,000 example? If you skipped that example, you better go back and read it now. To see this in action please visit the website, www.farmingwithoutthebank.com and click on How it Works.

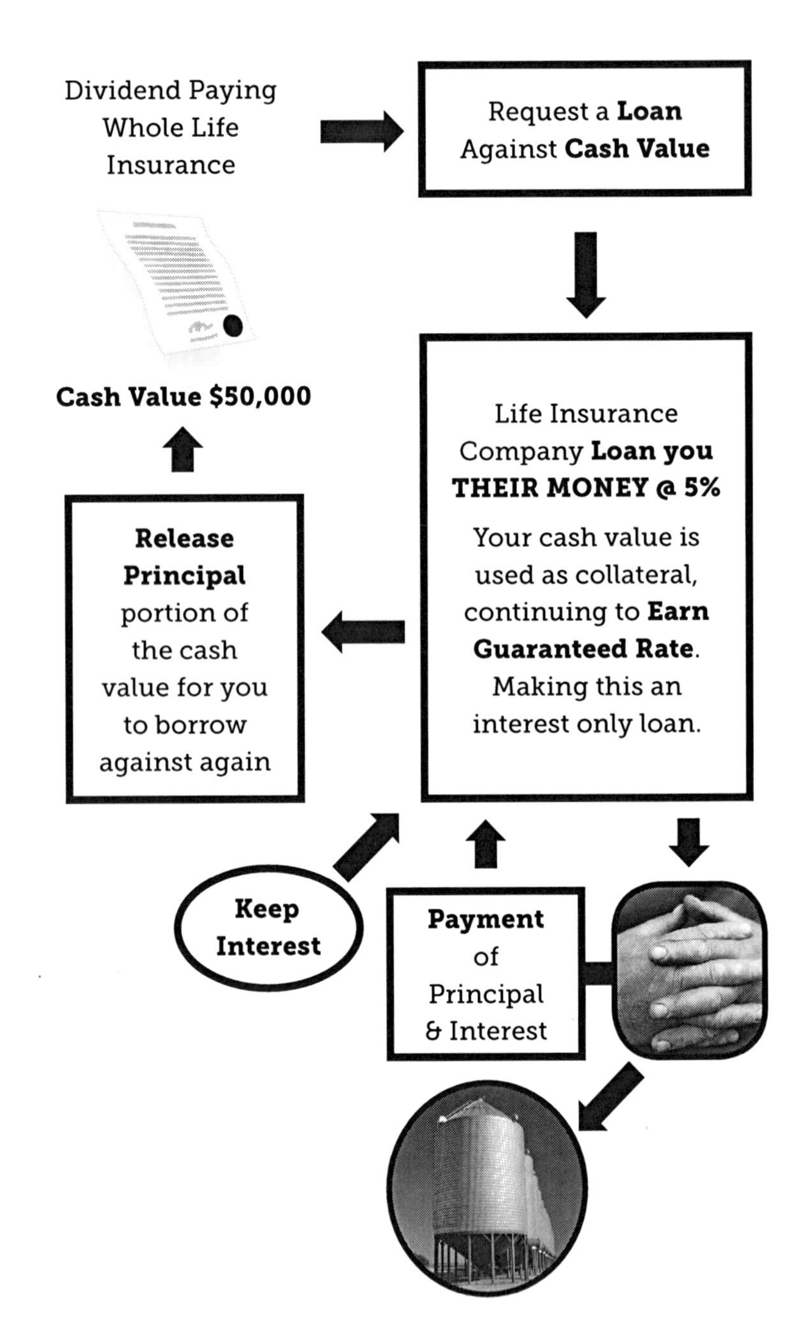

Dividend Paying Whole Life Insurance
Request a **Loan** Against **Cash Value**
Cash Value $50,000
Life Insurance Company **Loan you THEIR MONEY @ 5%**
Your cash value is used as collateral, continuing to **Earn Guaranteed Rate**. Making this an interest only loan.
Release Principal portion of the cash value for you to borrow against again
Keep Interest
Payment of Principal & Interest

Now stop and think about this: If you took money from a savings account and paid cash for your purchase, the account would drop by the amount you withdrew. Your account would pay you zero interest on the money you withdrew, because they can't pay you interest on money that is not there. That is the lost opportunity cost as we talked about before. Instead, using the whole life insurance this way allows you to borrow against the cash value to buy land or cattle **all while you continue to get paid interest on your money sitting in the cash value account**. In doing this, you continue to earn interest on your money and realize a rate of return on your land or cattle.

When you have a loan against your cash value, the life insurance company uses the death benefit as collateral to the loan. This is similar to using your land as collateral to your bank loan. The difference is you are no longer putting your assets and livelihood up as collateral for Mr. Banker; instead you are using your death benefit as collateral. Should you pass away with a loan against your cash value your beneficiaries will receive the death benefit minus the loan amount.

This is a very secure loan for you, because no one is going to take your assets should you have a couple of bad years farming and not be able to make payments. That is part of what makes this the perfect financial tool for you in good and bad years. This will get you through because you are in control. Being two to three years away from bankruptcy won't be such a huge worry.

Here is a simple chart so you can see how much control you can take back my having any debt in your control rather than the banks.

Collateral	Bank	Life Insurance
Home	✓	
Land	✓	
Equipment	✓	
Cattle	✓	
Vehicles	✓	
Death Benefit	✓	✓

Dividends

Along with cash value growth you are likely to get a dividend. These dividends can become quite large as the years go by, which is one factor that makes this a better tool than bank accounts or certificates of deposit. Dividends are not guaranteed, as indicated earlier, but they have been paid for over a hundred years.

Insurance companies are some of the largest and most conservative institutions out there, and you benefit from that. Over the last 150 years insurance companies have survived while banks collapsed. This conservative mentality benefits you as a policy owner because these companies are making money even when the economy is struggling. They do what they do best, while you do what you do best.

Death Benefit

Planning

I talked briefly about the death benefit being used as collateral to the loan on the policy. Here I want to touch on the other benefits of having a cash value loan and the added bonus of the death benefit. When you are looking at death benefit you need to consider the difference between a term life benefit and a whole life benefit.

TERM

With your term insurance your death benefit stays the same throughout the life of the policy and you give up use of those premium dollars. If you bought $500,000 of death benefit 20 years ago, that is still what you have today.

WHOLE LIFE

With whole life insurance your premium will stay the same while your cash value and death benefit will continue to increase throughout the years.

Death benefits are truly an added bonus when using whole life for financing. You have had access to the cash value for operating and upon death you have taken care of the family.

Should you have your entire operation financed through your life insurance policy, when you pass away your beneficiaries will be left with no farm debt and still have some death benefit. Upon your death any loan balance will always be paid off with the death benefit, leaving your family debt free. Excess death benefit will go to your beneficiaries income tax free.

Today's Young Farmers

Death benefit is more important than just collateral to a loan. You know as well as I, there are more young people coming back to farm today than ever before. Those young farmers are working on the farm without owning it until their mid-40's to early 50's. Parents and land owners are not holding on because they want to, rather because they have to hold on or they end up paying huge sums of money to taxes. It just doesn't make good tax sense to sell, even to the son or daughter.

This scenario does not leave young farmers in a better position. Instead, now middle aged children often have to secure money to buy equipment and the land they have farmed for years from siblings or at auction, which puts them in a position where they have to start all over again with millions of dollars of debt. This is where the death benefit comes in handy again, giving middle aged farmers some of the money they need to buy that land rather than forcing them to borrow it from the bank.

Key-Man Policy

If your kids are home farming with you, then you should be looking at this concept from another angle. You are able to purchase a policy on other people's lives, which is called a key-man policy. If your child is helping you farm and you depend on him/her, you are legally able to insure his/her life and he/she is legally able to insure your life.

If young farmers buy a life insurance policy on their parents, they would have the cash value to use while the parents are alive. When the parents pass away, they would then have the death benefit for any financial take over needs. As a farmer said to me, "I don't want my kids to have to buy the farm for the third time." This just makes sense, this farmer is not trying to give his kids a handout but how many times does one family have to buy a farm? The kids having a policy on dad's life may be able to ease the pain of taking over.

Parents can and should do the same on their child. Should parents have a policy on children and should the child pass first, parents then receive the death benefit and are able to hire someone to help them farm. In most cases, if it's a family operation, more acres are being farmed than one person can handle.

This type of scenario should also be used in partnerships. Many family operations exist where sibling are farming together or neighbors have joined forces operating as a legal corporation or partnership entity.

Preparing for the future is so important, and there is no better way to prepare than to have life insurance policies on each person in the operation. The partner who survives will need those added funds to keep the operation going.

Market Scam

Investments

Many of you are being guided to take some tax breaks or diversify by putting some money into 401Ks or IRAs. As this is an option given to you by investment consultants and accountants, it does come with restrictions and consequences you may not like or may not have even considered.

Many ask why they have not heard of this concept before. That answer may be found in a book Rick Beuter wrote titled *The Great Wall Street Retirement Scam.* It is the best little book I have read to explain the history of the markets. He tells us IRAs and 401Ks began in the mid 1970's and early 1980's. It was a push to have employees put money into the market. Prior to this, only the wealthy used the market. Average Americans saved in traditional savings accounts and purchased certificates of deposit (CDs). Employees were now forced to handle their own retirement accounts and work with brokers rather than their employers handling it for them, as they did prior to this change. During this process of moving average Americans into the market, the market soared and large rates of returns (ROR) were seen. These large rates of return drove the theory that the market was the place to be to make money.

In order to make this new system even more enticing, employees were given tax incentives to put their money in a retirement account tied to the market. Today most people have a 401K or IRA because it's all they know. Other options are not even presented. What you are seeing now is a very unstable market, and while you may not be making much, you continue to contribute because you hear of these great ROR's and how they are going to come back.

Rates of Return

Dwayne and Suzanne Burnell do an amazing job explaining rates of return in their book *Financial Independence In The 21st Century*. They explain in Chapter 2 that you will make more at 5% compounding than you will at an 8% rate in the market with a one-time loss of 20%. That is a one-time loss of 20% in ten years. Think of how many downturns the normal person goes through while they are invested. The authors also go on to explain that fees and taxes have not been figured into the overall average ROR investors talk so much about. After you figure your fees and taxes your true ROR is much lower than you were lead to believe.

Put the ROR's aside and take a look at what these products bring at retirement. You may get that tax break now as an incentive to put money in the market, but you are going to pay taxes when you take it out. A great example of this is seed versus harvest. If you went to the elevator to buy seed and they asked if you wanted to pay tax on the seed or the harvest, what would you do? You would pay on the seed because it would be a lot less than you would pay on the harvest. Yet, you are doing the exact opposite with your invested money; you are deciding to pay on the harvest just to get a tax incentive on the seed.

Taxes and Retirement

The tax break is very convincing when you hear you will be in a lower tax bracket at retirement, but is that the whole story? In order to be in a lower tax bracket you have to make less money per year OR taxes must go down.

Let's tackle this idea of making less. In order to make less you have to live on less, all this during retirement when you still have to eat, pay living expenses, buy vehicles and maybe go see the world. I have yet to meet anyone over retirement age who is living on less and enjoying that change in lifestyle. Remember Parkinson's Law? You don't want to downsize or change your standard of living once you are comfortable. Retirement should be comfortable not uncomfortable.

If you decide to rent out your land you will have a yearly income from land rent, plus social security, and possibly retirement accounts. This is all income you must claim. This will, in an ideal situation, keep you at the current income level you are at prior to retirement. If that is the case, your income stayed the same, your tax bracket stayed the same or higher, and your deductions decreased because you are no longer farming.

What this all means to you is your money from your 401K or IRA is likely going to be taxed at the same bracket or higher when you retire. In order to reduce taxes during retirement you may be forced to reduce lifestyle because you have to take less income from a retirement account just to reduce taxes.

There are also investment rules to keep in mind, such as having to withdraw money from your retirement accounts at age 70½. There is no avoiding the use of this money and thinking you will pass it on to heirs. The government wants its share of taxes and is going to do whatever it takes to get it. Making the 70½ rule was just one way to ensure taxes will be collected.

Whole life insurance allows you some tax benefits that other products do not offer. Tax deferred growth on the guaranteed interest rate of the policy is one of those tax advantages. You are getting it tax deferred because you are paying tax before you ever pay premium. Dividends are a tax-free event as well. You paid extra premium based on projections and are getting a return of money which is tax free.

By borrowing against your cash value to supplement retirement you do just that. You decrease your tax burden because the money used to supplement your income is not taxable due to the fact that it's technically a loan.

Diversification

You may be the farmer putting money away for retirement to keep your money diversified or get the immediate tax benefits. Most accountants will say it's your best option to get some tax breaks for the year. However, there are several reasons why these 401k and IRA accounts may not be your best option.

1. You are not truly diversified. Every day your farm is at risk due to weather and circumstances out of your control. If you have money in the market, it too is at risk and out of your control. There is no elimination of risk or guarantee in anything you have when you use this method. There is not even insurance available for a market crash. At least you are able to insure your crops and cattle against loss.
2. Your money is tied up in that account, while you could be using it to farm,until you are 59½ or you pay a 10% penalty to access it early. There are no penalties to access cash value like there are in your 401K's and IRA's. Even a CD is locked up until the term is up. This is where storage of money versus utilization of money becomes very important.

3. When using 401K or IRA money to supplement your income during retirement it shows as "income" on which you will pay taxes.
4. You are forced to start taking money from investments when you are 70½ regardless if you need it or not. Again, this forces income and taxes on you for the benefit of them.

As a farmer, your need for financing options is much greater today than your need for retirement. If you are like most farmers I work with, you will have land to lease out at retirement supplying you with yearly income. Cash value gives you access to the money today for financing and will double later in life to supplement retirement with tax benefits.

POLICY UNDERSTANDING

Illustrations

Before you go any further it is important to understand the basics of how to read a whole life insurance illustration. So many people are unfamiliar with whole life insurance; getting thrown into the numbers of an illustration can be like reading a foreign language.

An illustration is nothing more than a projection or outlook of what your policy will look like in the future.

Below you will see there are several columns on a whole life insurance illustration, so let's go over them in detail.

Guaranteed and Non-Guaranteed Assumptions Columns

Guaranteed side means by contract the company will guarantee you will have the illustrated amount of cash value and death benefit they have projected.

Non-Guaranteed Assumptions side figures in

anticipated dividends. Dividends are not something companies can guarantee, but remember they have successfully paid dividends for over a century.

Contract Premium

Contract Premium is the amount of premium you have agreed to pay each year for the life of the contract. This will never increase, and to decrease it there are restrictions. It is very important to make sure you are comfortable with this amount from the beginning, because you can only decrease, never increase, premium. Notice too how premium appears under the guaranteed column.

Net Cash Value

Net cash value is the amount of money you are able to borrow against or receive should you surrender your policy. You will notice there are two columns for cash value. The Non-Guaranteed side includes the dividends, whereas the Guaranteed side does not.

Death Benefit

Death benefit is the amount of money your beneficiaries will receive when you die. Again there are two columns: Non-Guaranteed includes the death benefit with dividends being paid. Guaranteed does not.

Cum(ulative) Premium

Cumulative Premium is the total amount of premium you have paid to the life insurance company up to that particular year.

Annual Dividend

Annual Dividend is your anticipated yearly share of the profits from the company since you are part owner.

Increase in Net Cash Value

Increase in Net Cash Value is the amount your net cash value increased at year end after premium and dividends (non-guaranteed) were paid.

Throughout the rest of the book when you see case studies, the numbers used will be from the Non-Guaranteed side. With the company's long history of paying dividends, there is no reason to assume dividends will not be paid. Keep this page handy when moving forward, as you may want to reference these terms again.

		Guaranteed			Non-Guaranteed Assumptions 100% of Current Dividend Scale					
Age	Year	Contract Premium	Net Cash Value	Death Benefit	Contract Premium	Cum Premium	Annual Dividend	Increase in Net Cash Value	Net Cash Value	Death Benefit
36	1	12,000	7,818	841,361	12,000	12,000	130	7,948	7,948	842,075
37	2	12,000	15,946	882,662	12,000	24,000	173	8,306	16,254	884,288
38	3	12,000	26,091	922,387	12,000	36,000	224	10,381	26,635	925,149
39	4	12,000	37,740	960,598	12,000	48,000	275	11,946	38,580	964,705
40	5	12,000	49,844	997,360	12,000	60,000	332	12,469	51,049	1,003,027
41	6	12,000	62,406	1,032,727	12,000	72,000	394	13,003	64,052	1,040,175
42	7	12,000	75,432	1,066,757	12,000	84,000	464	13,554	77,606	1,076,227
43	8	12,000	88,929	1,099,508	12,000	96,000	544	14,126	91,732	1,111,260
44	9	12,000	102,896	1,131,035	12,000	108,000	635	14,710	106,442	1,145,351
45	10	12,000	117,344	1,161,393	12,000	120,000	732	15,315	121,757	1,178,554

Young VS Old

So many times I hear, "If only I was younger. I am too old to start using this, but it would be good for those young kids." This is a big misconception. Yes . . . age is a factor, but it is not the end all. Regardless if you are 35 or 65 years old you still use money on a daily basis, and this should be something you at least consider. Furthermore, the bank doesn't turn you away from making deposits as you get older.

In order to put this misconception to rest, below you will see an illustration using the same amount of premium dollars each year for the next ten years, with the only difference being age: a 35 year old and 55 year old.

To compare, you really need to look at two columns: Net Cash Value and Death Benefit.

Cash Value

In the cash value column take a look at year one. You can see the 55 year old actually has more cash value than the

35 year old. In fact, the 55 year old has more money in cash value all the way through year nine, at which point the 35 year old starts to somewhat outperform. The 55 year old continues to perform equally as well with their cash value all the way to year fifteen.

Another thing to look at is the "break even" point. This is the point at which the cumulative premium equals the net cash value. In BOTH cases that happens at year ten.

The reason for this equal performance because you are older you have less time to accumulate money, giving you an opportunity to put more money into the policy early on. The younger you are the longer you have, which makes the company look long term at how much you will accumulate in cash value compared to your death benefit. All these factors are taken into consideration when building a policy for the Infinite Banking Concept®.

Death Benefit

When looking at the death benefits, it's a much different picture. At the age of 55 it costs more for insurance, so you have a far less death benefit than you would have at 35. However, just fifteen years later the death benefit at seventy is still nearly a half million dollars.

In either case, your family will be taken care of because you were building a financial system and using the living benefits to finance the farm and supplement retirement tax-free. If you had a term policy you would not have had the cash value to use during this time period, may not

Illustration 1 Young vs Old Comparison

35 Year old Year	36 Year old Age	Premium	Cummulative Premium	Increase in Cash Value	Net Cash Value	Death Benefit
1	36	12,000	12,000	7,952	7,952	856,464
2	37	12,000	24,000	8,311	16,263	898,700
3	38	12,000	36,000	10,467	26,730	939,585
4	39	12,000	48,000	12,089	38,819	979,164
5	40	12,000	60,000	12,618	51,437	1,017,511
6	41	12,000	72,000	13,158	64,595	1,054,684
7	42	12,000	84,000	13,715	78,310	1,090,761
8	43	12,000	96,000	14,293	92,603	1,125,822
9	44	12,000	108,000	14,882	107,485	1,159,942
10	45	12,000	120,000	15,494	122,978	1,193,176

55 Year old Year	56 Year old Age	Premium	Cummulative Premium	Increase in Cash Value	Net Cash Value	Death Benefit
1	56	12,000	12,000	8,490	8,490	373,905
2	57	12,000	24,000	9,122	17,613	395,559
3	58	12,000	36,000	11,511	29,124	416,724
4	59	12,000	48,000	11,963	41,087	437,423
5	60	12,000	60,000	12,430	53,517	457,702
6	61	12,000	72,000	12,904	66,420	477,616
7	62	12,000	84,000	13,377	79,797	497,218
8	63	12,000	96,000	13,858	93,655	516,565
9	64	12,000	108,000	14,347	108,002	535,705
10	65	12,000	120,000	14,852	122,854	554,661

have coverage after age 65 or 70, and the death benefit would not have increased over time. Regardless of what financial gurus tend to say, it is important to have coverage and be debt free when you are over 65. Life does not end; you continue to use money. If you are the typical farmer, you don't stop farming, which always leaves you with a need for financing.

As you can see, cash value still performs just as well at 55 as it does at 35. Age is a factor that attributes to a lower death benefit, but it is not impossible to use as a banking tool as one might think.

Policies on Young Children

Thinking that a policy will perform better on children is another misconception. First you have the obstacle of state guidelines to write insurance on children. In North Dakota the child can only have half the death benefit the parents have. To obtain a copy of your state guidelines, check with your local agent or state's insurance department. If you are the parent and have $500,000 death benefit, your child can only have death benefit of $250,000. It does not take much premium to get a child's death benefit to $250,000. So, when looking at it as a financial tool, a child's policy will not build cash value any faster than an adult's cash value. In fact, it will be much slower, because you can't fund it with any significant amount of money.

This does not mean you should not be starting a policy on your child. In fact, you should absolutely consider starting a policy on your child. I see far too many young people stuck into programs like FSA because they are young and interest rates are good. These programs are great for helping them get started, but the lack of control over their finances is mind boggling. Giving your child a head start will alleviate the hassle of such programs.

Most kids growing up on farms have an opportunity to make some money through grain sales or cattle of their own. In our family, each child got a cow when we were little, and any sales from that cow's offspring were ours. If your family does this, the money from sales could go to fund a policy and your kids could be in a position to start their own policy. This would give them a leg up to build cash value for a later time when they will need it.

Teaching our kids to use this tool will change the way farming is done for generations to come. Just as you teach your kids to farm, you should be teaching them how to correctly finance the farm. They would learn at a young age to borrow against their policy for purchases and pay themselves back PLUS interest.

Case Studies

Operation Financing

You have all the information you need on being an honest banker: why compounding interest is key, the tax benefits of using this tool, and why we use whole life insurance. Now you need to see it implemented.

You can use this concept on many levels from vehicle purchases to operating expenses to financing land/cattle/equipment purchases. I will discuss all these areas in the next few chapters.

Before you go any further, I need you to keep in mind nothing happens overnight and there is no magic wand that will make the banks disappear quickly. Your crop does not come up overnight, nor does money grow that fast.

From this point forward there are case studies to show you how this works. Use your imagination when looking at these cases. The numbers are all relative to the size of the operation. Hundreds of thousands of dollars can be a lot of money to some and only a fair amount to others. Your expenses are relative to your operation, as is the case for most farmers. If the premium looks too high for you, then cut it in half or remove a zero. The numbers still work! If the premium is not high enough, increase the amount for purposes of illustration. A certified practitioner can customize an illustration to your specifications and circumstances.

In the following case study we are going to use Don and Anita, who are 45 years old, and farming with their son Lane, who is 20 years old. Lane is living at home during the summer and going to college the rest of the year for farm management, with intentions of taking over the family farm someday. Lane is a hard working farm kid who comes home every weekend he can to help out.

Case Study 1: Financing Operating Expenses

(*all numbers have been rounded*)

Don and Lane are farming around 3,000 acres of grain and have a yearly operating budget of $500,000, which they estimate will increase by 2% yearly. Don figures he can fund a policy on himself for $100,000 in premium each year for the next 13 years. At year 14 and later, his premium drops to $50,000, then again to $25,000, and again to $22,100 until the age of 100. This change in premium happens because we are trying to keep the policy within the modified endowment contract (MEC) guidelines to maximize tax benefits. (For more detailed information on the MEC please see pages 37–38 of *Becoming Your Own Banker*.)

By setting this policy up for banking, Don has the availability to a cash value of $74,075 the first year but his operating expense is $500,000. As you can see this does not allow Don the ability to walk away from Mr. Banker just yet. Don will borrow against his cash value available but will still need to go to the bank and borrow the additional $425,925 he needs to operate for the year.

Each year Don pays back his operating loans in full. At the end of the year Don will pay off his operation note to the bank and his cash value of the $74,075. As you can see Don has not done anything differently—he still owed $500,000 at the end of the year. The only thing that changed is who he made those payments to. Instead of it all going to the bank, some of it came back to him. Now that he has paid back his cash value in full, he has that $74,075 available to borrow against again for next year's operating loans.

The next year comes (year two) and he does the same thing. He pays his premium of $100,000 and his cash value increased by $77,340 giving him a total of $151,415 to borrow against for year two's operating expense. This year he has to borrow even less from the bank, $358,585. At the end of year two he does the same thing as last year—he pays the bank his portion and the other back to his policy loan, again, leaving him with $151,415 to borrow against for year three's operating expense.

As you can see by the illustration, Don continues this process and by year six his operating loan is LESS than

Case Study 1A Don & Anita Financing Operating Expenses

Age	Year	Premium	Cumulative Net Premium Outlay	Increase in Net Cash Value	Net Cash Value	Yearly Operating Expense	Local Bank Loan or Excess Cash Value	Net Death Benefit
46	1	100,000	100,000	74,075	74,075	500,000	-425,925	2,634,243
47	2	100,000	200,000	77,340	151,415	510,000	-358,585	2,903,486
48	3	100,000	300,000	99,206	250,621	520,200	-269,579	3,164,963
49	4	100,000	400,000	104,525	355,146	530,604	-175,458	3,418,790
50	5	100,000	500,000	108,963	464,108	541,216	-77,108	3,665,464
51	6	100,000	600,000	113,562	577,670	552,040	25,630	3,905,370
52	7	100,000	700,000	118,151	695,821	563,081	132,740	4,138,885
53	8	100,000	800,000	122,929	818,750	574,343	244,407	4,366,583
54	9	100,000	900,000	127,765	946,516	585,830	360,686	4,588,954
55	10	100,000	1,000,000	132,645	1,079,161	597,546	481,615	4,806,719
56	11	100,000	1,100,000	137,660	1,216,820	609,497	607,323	5,020,384
57	12	100,000	1,200,000	142,898	1,359,718	621,687	738,031	5,230,391
58	13	100,000	1,300,000	148,385	1,508,104	634,121	873,983	5,436,822
59	14	50,000	1,350,000	105,218	1,613,321	646,803	966,518	5,524,231
60	15	25,000	1,375,000	84,565	1,697,887	659,739	1,038,148	5,555,441
61	16	25,000	1,400,000	87,469	1,785,355	672,934	1,112,421	5,588,502
62	17	25,000	1,425,000	90,253	1,875,608	686,393	1,189,215	5,623,721
63	18	25,000	1,450,000	92,994	1,968,602	700,121	1,268,481	5,661,374
64	19	25,000	1,475,000	95,861	2,064,462	714,123	1,350,339	5,701,763
65	20	25,000	1,500,000	98,949	2,163,412	0	2,163,412	5,744,827
66	21	22,100	1,522,100	101,896	2,265,307	0	2,265,307	4,789,889
67	22	22,100	1,544,200	105,058	2,370,366	0	2,370,366	4,836,655
68	23	22,100	1,566,300	108,387	2,478,753	0	2,478,753	4,885,084
69	24	22,100	1,588,400	111,916	2,590,669	0	2,590,669	4,935,071
70	25	22,100	1,610,500	115,407	2,706,076	0	2,706,076	4,986,644
71	26	22,100	1,632,600	118,891	2,824,967	0	2,824,967	5,040,174
72	27	22,100	1,654,700	122,074	2,947,041	0	2,947,041	5,096,109
73	28	22,100	1,676,800	125,097	3,072,138	0	3,072,138	5,154,856
74	29	22,100	1,698,900	128,281	3,200,419	0	3,200,419	5,216,260
75	30	22,100	1,721,000	131,660	3,332,079	0	3,332,079	5,280,250
76	31	22,100	1,743,100	134,908	3,466,987	0	3,466,987	5,346,605
77	32	22,100	1,765,200	138,018	3,605,005	0	3,605,005	5,415,374
78	33	22,100	1,787,300	140,932	3,745,936	0	3,745,936	5,487,005
79	34	22,100	1,809,400	143,571	3,889,508	0	3,889,508	5,561,957
80	35	22,100	1,831,500	145,950	4,035,458	0	4,035,458	5,640,422

his cash value available. This is the year he does not have to go the bank to borrow money for operating. He is able to borrow the full amount against his cash value. At this point Don has an excess of $25,630 in cash value after he has borrowed the full amount needed for operating.

Look at each year following—there is excess cash value, even though operating expenses increase by 2% each year. This cash value continues to grow larger and larger giving Don options for what he can do with the extra money. These options will be looked at in subsequent chapters.

Multiple Loans

You have seen how the policy can be used for one specific loan and many purposes throughout life. If you recall earlier in the book there was talk about not compartmentalizing your money. You will see just what was being referenced in this case study.

We visited about Lane and the fact that he was going to take over the farm when Don retired. Remember too, Lane was young when Don started his policy, and Don and Anita made sure Lane knew how these policies worked and had him start his own.

Case Study 2: Lane's Policy with Multiple Loans (*all numbers have been rounded*)

Lane started his policy at age 20 while he was in college. He didn't have a lot of extra cash but knew he needed to get something started to begin building cash value. He decided he could manage a $12,000 premium, because Don and Anita were paying him to help on the farm.

Lane did not borrow against his cash value until he was done with college. After he completed college at age twenty-three he decided to borrow $30,000[1] against his cash value to pay off his student loans. He did not want to drag the payment out for too long knowing he would need it for farming, so he decided to pay it back over the next five years, making a $6,000/year payment. The reason you see a loan of $24,000 and not $30,000, is because it is showing the $6,000 pay back for that year.

By the time Lane was 27, some of the neighbor's land came up for sale and he wanted it. His student loans were paid back by this time, and he had enough cash value to borrow against for the purchase of this land. Not only could he farm some of this land, he could also put a house on a piece of it for his own family one day. At year eight he borrowed against $80,000[2] and decided to pay this loan back over the next seven years. This would make his yearly payment $11,429.00

Since he had been dating Sarah for a while, he figured it was time to get married. Lane and Sarah married at age 34 and built a house on that piece of land he bought. Knowing he would need the majority of his cash value for operating soon, he decided to only borrow against $50,000[3] for a down payment on the house. Borrowing $50,000 would make the house payment affordable, and he knew he could be flexible paying the policy loan back. They decided to make the policy loan a fifteen year payback term with a payment of $3,333 a year. This gave them room in their budget for home furnishings.

Lane and Sarah took a couple years of no policy loans just to have some extra money in reserves. Then Lane decided it was time to set the wheels in motion and start operating a portion of his dad's farm. Lane had always helped out financially here and there, but this was the first year he was going to have a big operating loan and buy the majority of the seed and fertilizer himself. Lane borrowed against $250,000[4] to use for operating. As you can see in the Case Study 2A table (page 73), Lane did not have enough cash value and needed to go to the bank for the extra $36,888[5]. You can see he also needed a little extra from the bank the following year, but by the third year he had enough cash value to use only that.

He continued using the policy for operating until he was 40, at which point Don was now 65 and ready for Lane and Sarah to farm it all so he could retire. Lane and Sarah did not yet own the land but needed operating money to farm all of it, which meant a very large loan of $500,000[6]. Lane was just a little short on what he needed, so again he had to go to the bank to borrow $177,200[7] that year. He also had to continue using the bank to fund a portion of his operating loan for the next five years.

Lane operated on $500,000 until his dad passed away at age 79, at which point Lane was 54 and having to buy land and equipment from his mother. He needed access to a million dollars[8] in order to buy the land and operate that year. At this point he again only had to borrow $157,773[9] from the bank to take care of the estate. What a great relief it was to have access to available cash and not worry about having to borrow a million dollars at age 54 and feel like he was starting over.

Look at what Lane has done with this policy over the course of 30 years. He has paid for college, bought land, operated the farm, built a house and bought out his parents' estate. Lane was grateful that Don and Anita shared this financial strategy with him; it truly put him a step above the rest of his peers who were taking over their family farms. Lane did not have to fall victim to FSA or other programs and only had to utilize the banks a few times.

One thing we have not talked about is Lane's death benefit. Take a look, at any age, he has well enough to take care of his family should something happen to him.

This is a great illustration of how important it is to teach your young adults this system. The sooner they get started the better.

Another thing consider here is Lane did not have a key-man policy on his father. As shown on page 47 these policies are almost necessary for a family operation. Had Lane taken a policy out on Don than upon Don's death Lane would have had a death benefit to use to purchase the estate and possibly avoided having to borrow money from his policy or the bank. Again, be sure to think of these things with your own operation if you are working together with family or partners.

Case Study 2A Lane's Policy with Multiple Loans

Age	Year	Premium	Cumulative Net Premium Outlay	Net Cash Value	Yearly Operating Expense	Local Bank Loan or Excess Cash Value	Net Death Benefit
20	1	12,000	12,000	7,986		7,986	1,026,744
21	2	12,000	24,000	16,326		16,326	1,101,601
22	3	12,000	36,000	25,030		25,030	1,174,272
23	4	12,000	48,000	35,230	[1]24,000	11,230	1,244,854
24	5	12,000	60,000	47,899	18,000	29,899	1,313,386
25	6	12,000	72,000	61,128	12,000	49,128	1,380,024
26	7	12,000	84,000	74,924	6,000	68,924	1,444,793
27	8	12,000	96,000	89,336	[2]68,571	20,765	1,507,697
28	9	12,000	108,000	104,427	57,143	47,284	1,568,741
29	10	12,000	120,000	120,214	45,714	74,500	1,627,942
30	11	12,000	132,000	136,742	34,286	102,456	1,685,284
31	12	12,000	144,000	154,045	22,857	131,188	1,740,885
32	13	12,000	156,000	172,139	11,429	160,710	1,794,775
33	14	12,000	168,000	191,060	0	191,060	1,847,137
34	15	12,000	180,000	210,846	[3]46,667	164,179	1,897,960
35	16	12,000	192,000	231,515	43,333	188,182	1,947,392
36	17	12,000	204,000	253,112	[4]290,000	[5]-36,888	1,995,458
37	18	12,000	216,000	275,671	286,667	-10,996	2,042,306
38	19	12,000	228,000	299,215	283,333	15,882	2,088,098
39	20	12,000	240,000	323,823	280,000	43,823	2,132,818
40	21	12,000	252,000	349,467	[6]526,667	[7]-177,200	2,176,433
41	22	12,000	264,000	376,210	523,333	-147,123	2,219,211
42	23	12,000	276,000	404,069	520,000	-115,931	2,261,130
43	24	12,000	288,000	433,066	516,667	-83,601	2,302,389
44	25	12,000	300,000	463,244	513,333	-50,089	2,343,019
45	26	12,000	312,000	494,653	510,000	-15,347	2,383,180
46	27	12,000	324,000	527,342	506,667	20,675	2,422,821
47	28	12,000	336,000	561,407	503,333	58,074	2,461,813
48	29	12,000	348,000	596,925	500,000	96,925	2,500,005
49	30	12,000	360,000	633,926	500,000	133,926	2,537,531
50	31	12,000	371,646	672,486	500,000	172,486	2,274,559
51	32	12,000	383,292	712,555	500,000	212,555	2,311,203
52	33	12,000	394,938	754,208	500,000	254,208	2,347,664
53	34	12,000	406,584	797,441	500,000	297,441	2,384,020
54	35	12,000	418,230	842,227	[8]1,000,000	[9]-157,773	2,420,509

Supplementing Retirement

Let's go back to Don and his retirement years. Don has his loans paid back in full and has stopped farming at age 65. He has $2,163,412 in cash value to supplement his income (as shown in the illustration on page 68), tax free. In addition to the tax benefits, this cash value will supplement Don's and Anita's income through loans without having to be paid back.

This cash value of over $2 million will give Don and Anita all kinds of options. They may decide to borrow against the cash value to pay a premium of $22,100 each year. If they rented out their land they may not need to do that and can instead lend money to Lane if he needed it. Or they may just live it up and use it all traveling the world. Their options are endless.

Don still has a premium of $22,100 to pay each year from age 66 to 100 which he could stop paying. He decides to continue paying the premium, because he knows his cash value will increase substantially by doing this and he has the income to do so.

This increase in cash value is one of the reasons you do not want to cancel your policy when you are in retirement. If you want access to the cash value, borrow against it and do not cancel the policy. If Don would cancel this policy he would give up access to extra money each year from 66 on. The only people who would advise this are those who don't understand what is happening in the cash value portion of the policy. They are solely looking at premium in the form of a payment rather than a deposit. As I said earlier, it truly is acting as a deposit within a policy that is set up correctly for financing.

When you retire, the goal is to maintain the standard of living you currently have. Remember Parkinson's Law? You should be able to travel, spend some time with and spoil the grandkids rotten, and enjoy your time. I know all this may be hard for you since farmers don't stop working, but if my grandparents can stop farming, so can you.

Regardless of what you plan to do, you are going to need money to maintain your lifestyle, and your goal is to do this with as little tax implications as possible.

Case Study 3: Supplementing Retirement

Let's look at Don's policy again, but this time for supplementing retirement income for him and Anita.

At age 66 Don decides to start borrowing $75,000 a year against his cash value to supplement their income. Remember, Lane took over, so Don has no farm income to pay back this loan or interest each year and that is ok. Don is not expected to pay back the loan during retirement; his death benefit will take care of the loan amount when he dies.

With $2 million to borrow against, he figured $75,000 a year would be a comfortable amount. You can see at the age of 79 he still has $2.3 million of cash value and $4 million of death benefit. If Don decided to live it up before he passed away at 79 and borrow all $2.3 million, his heirs would still get $1.7 million of death benefit.

What you can see here is that if Don lived to 85 he would be able to continue taking this $75,000. What you don't see is that it will be until he is 100 years old before he would run out of cash value. All that time, he can be taking this loan and never paying the loan or interest back to the company. Again, that is ok. Whatever is taken out and not paid back will be deducted from the death benefit prior to payout.

Should Don have bad family genes and expect to pass away earlier then, of course, he would have the ability to take far more than $75,000/year. The amount borrowed each year is not predetermined. Don is able to be flexible and take more or less as he needs it. The only restriction Don has is the amount which he can borrow against. He can borrow up to the full amount of his cash value. Once the equivalent of the cash value has been collateralized or borrowed against, he has reached the limit and can no longer borrow without repaying at least part of the loan.

To repeat again, this money comes to Don tax-free because he borrows against it. Few tools you use for retirement come to you tax free. You have to pay tax on those IRAs and 401Ks all the way down to your land. Most farmers are not even selling land because of the tax implications. If your land and retirement income was all taxable income that does not put you in a lower tax bracket, rather it keeps you where you are. However, borrowing against cash value may surely make a difference in your tax table, since it is not taxable income.

In addition, this policy provided Anita with a death benefit upon Don's death (if she out lived him). In our scenario of Don's passing at age 79, Anita would have received a death benefit of $2.9 million. Remember they borrowed against $75,000 a year for 14 years before he passed. So the death benefit minus the $1,050,000 loan still leaves Anita well taken care of.

Case Study 3A Supplementing Don's Retirement

Age	Year	Premium	Loan Amount	Net Cash Value	Net Death Benefit
66	21	22,100	-75,000	2,186,557	4,711,139
67	22	22,100	-75,000	2,208,928	4,675,218
68	23	22,100	-75,000	2,230,494	4,636,824
69	24	22,100	-75,000	2,251,247	4,595,649
70	25	22,100	-75,000	2,270,933	4,551,500
71	26	22,100	-75,000	2,289,317	4,504,524
72	27	22,100	-75,000	2,305,858	4,454,926
73	28	22,100	-75,000	2,320,146	4,402,864
74	29	22,100	-75,000	2,332,077	4,347,918
75	30	22,100	-75,000	2,341,570	4,289,741
76	31	22,100	-75,000	2,348,202	4,227,820
77	32	22,100	-75,000	2,351,531	4,161,900
78	33	22,100	-75,000	2,351,039	4,092,108
79	34	22,100	-75,000	2,346,116	4,018,565
80	35	22,100	-75,000	2,336,146	3,941,110
81	36	22,100	-75,000	2,320,521	3,859,505
82	37	22,100	-75,000	2,298,897	3,773,429
83	38	22,100	-75,000	2,271,030	3,682,249
84	39	22,100	-75,000	2,236,300	3,585,567
85	40	22,100	-75,000	2,194,103	3,483,089

Building the Family Bank

At some point Lane and Sarah have a child, Erica. Teaching and molding that young person into a well-rounded human being is a parent's primary job.

As we have already discussed, taking over a family operation is not easy, and we see how lucky Lane was to have access to cash. Having to borrow money late in life to buy land is frustrating.

Lane already knows why policy funding is essential, and he wants to be sure Erica has something set up for her as well. Teaching her how to use this concept and process may very well be the most important thing Lane and Sarah do. She can learn to farm but if she can't get the money to farm, the knowledge goes unused.

As new parents and young farmers they do not have an excess amount of income. However, Don and Anita are in a better position to fund a policy for Erica because they do have some disposable income. Mr. Nelson Nash talks about this family process and how it may have to skip the first generation, but if a family started the grandkids off correctly and taught them, they would soon have a family legacy much like the Rothschild's. Lane had to start his own, but now Don and Anita can continue this tradition for their grandkids.

Case Study 4: Grandchild Erica's Policy

This is an example of what it would look like to have Don and Anita start a policy on Erica when she is eight years old for just $100/month. In this case Don and Anita are the owners and Erica would be the insured. As owners of the policy Don and Anita have access to the cash value and pay premiums until they turn ownership over to someone else. While Erica is a minor, Don and Anita can decide to keep ownership, turn ownership over to the Lane and Sarah or make Lane and Sarah contingent owners should they pass away. When Erica reaches adulthood, Don and Anita can then turn ownership of the policy over to Erica. An important note to remember is, whoever takes over ownership of the policy has access to the cash value and assumes premium payments.

As you can see in the Case Study 4A (page 82), this is a little policy and cash value is minimal in year one, but at Erica's age that is okay. They are looking years down the road when Erica is a teen and is in the need of some bigger priced items where she will have the cash value she needs and use the policy.

Erica turns sixteen and needs a car. Don and Anita decide to lend some of the money from the cash value of her policy. The great lesson comes when they sit down with her and work out payback terms. They all decided together she will pay this loan back in three years because she will want to use that money again for college.

She decides to go to college at age eighteen, and at that time she would have $9,605.00 of cash value to borrow against to pay for or supplement her college tuition, depending on the college she attends. Now Erica needs to decide if she will have a job while she attends college so she can pay some of that loan back each month. Repaying just means there will be more money available for her to borrow against next semester or next year.

At this point Erica has used her policy twice. Don and Anita, along with Lane and Sarah, are teaching her how to be an honest banker. Teaching your kids to be honest stewards of their money is one of the greatest pieces of knowledge you can share as a parent or grandparent.

By age 30, Erica is ready to start purchasing some assets of her own, such as cattle, and has $29,662.00 of cash value against which to borrow. Now that may not seem like a lot to get started, but everyone has to start somewhere. This is money she didn't have to borrow from the bank and worry about paying back, giving her a little freedom to make some mistakes or have a bad year. Remember the young lady I spoke of earlier in the book with $20,000 who was limited to how many cattle she could purchase? Erica won't have anybody tell her what and how much she can buy.

Wouldn't you know it, Erica had unexpected veterinarian fees because of an illness. Her cattle expenses increased and paying back loans in full was not an option. Having this system in place gave her the freedom to finance these cattle purchases herself without the worry of paying back banks or being questioned as to why her veterinarian bill was too high.

When a family banking system is started, the entire family, young and old, will understand how to operate on their own and become financially independent. Family banking systems are more than just the owner using the policy. It is about Don and Anita having the ability to lend excess cash value to Lane and Sarah if they need it, and Erica knowing how to borrow money efficiently rather than through banks. It's about generations of doing the same and creating a dynasty and keeping the farm in tack.

Use the story of the Vanderbilt's and the Rothschild's as an example. In 1877 Cornelius Vanderbilt died the richest man in the United States leaving an estate of $105 million to his heirs. By 1973 there was not one Vanderbilt left that was a millionaire. Yet the Rothschild's family has experienced something different. When Amschel Rothschild died in 1812 he had a banking dynasty set up for his heirs. Prior to passing, he taught his five sons conservative money management. They borrowed from a family banking system and were taught to pay back each time they borrowed from it. Each family member passes this knowledge down to their children, and, in order to use this family banking system, mandatory attendance of annual meetings is required. Should they not attend this meeting, they are not allowed to borrow from the family banking system. Today the Rothschild's dynasty is in the trillions.

Research shows that family money rarely survives the transfer for long, with 70% evaporated by the end of the second generation. By the end of the third generation, 90% will have evaporated. Think about what you can teach your family for future success. Knowing how to farm is just one part of the equation. One cannot farm without money, and you do not want your family farm gone by the third generation.

Case Study 4A Grandchild Erica's Policy

Age	Year	Premium	Cumulative Net Premium Outlay	Increase in Net Cash Value	Net Cash Value	Net Death Benefit
9	1	1,200	1,200	595	595	127,604
10	2	1,200	2,400	622	1,217	135,985
11	3	1,200	3,600	649	1,866	144,053
12	4	1,200	4,800	677	2,543	151,823
13	5	1,200	6,000	988	3,431	159,307
14	6	1,200	7,200	1,156	4,684	166,523
15	7	1,200	8,400	1,185	5,869	173,490
16	8	1,200	9,600	1,212	7,082	180,226
17	9	1,200	10,800	1,244	8,326	186,743
18	10	1,200	12,000	1,279	9,605	193,061
19	11	1,200	13,200	1,324	10,929	199,184
20	12	1,200	14,400	1,374	12,304	205,121
21	13	1,200	15,600	1,431	13,735	210,876
22	14	1,200	16,800	1,494	15,229	216,456
23	15	1,200	18,000	1,556	16,784	221,864
24	16	1,200	19,200	1,622	18,406	227,104
25	17	1,200	20,400	1,688	20,095	232,182
26	18	1,200	21,600	1,759	20,854	237,110
27	19	1,200	22,800	1,828	23,682	241,885
28	20	1,200	24,000	1,903	25,585	246,520
29	21	1,200	25,200	1,992	27,576	251,011
30	22	1,200	26,400	2,085	29,662	255,360
31	23	1,200	27,600	2,181	31,843	259,570
32	24	1,200	28,800	2,283	34,126	263,645
33	25	1,200	30,000	2,386	36,511	267,596
34	26	1,200	31,200	2,493	39,004	271,426
35	27	1,200	32,400	2,597	41,601	275,132
36	28	1,200	33,600	2,710	44,311	278,724
37	29	1,200	34,800	2,818	47,129	282,214
38	30	1,200	36,000	2,930	50,059	285,598

Policy Funding and Comparisons

If you are like most people who come to see me, you get the concept and are now excited to become your own banker. You will have two reactions.

1 "I can't afford this right now." This is the most common reaction I see, and it is typically the most incorrect reaction.

When times were good, and you didn't need this concept then either because you had cash. When times are bad I hear there is not enough money to start. Nelson Nash figured this concept out in the 80's when he got stuck with a $500,000 land loan at 25% interest. There is never going to be a perfect time, you either give the banks interest and control or pay cash and give up the opportunity to make money and lose access to capital. It boils down to one thing, if you want to get started badly enough you will find a way.

2 You want to see how fast you can get away from the banking system by running all sorts of scenarios. What you will find is proof that more is better when you apply compounding interest and dividends. Here is where you don't want to be a Wal-Mart shopper.

Here you will find illustration comparisons showing various rates of return. Each policy was run on the same person but funded differently at $10,000, $20,000, $40,000 and $70,000 a year for 20 years.

As you can see, the rate of return is 3.12% and 3.14% for the $10k and $20K policies. Then you see a 3.38% and 3.39% for the $40K and $70K because you have more money working for you over the same time period.

The "break even" point, the point where total premium paid equals cash value, is two years sooner on the bigger policies as well. Even the increase in cash value each year performs better with the high premiums.

Higher premiums mean bigger dividends. The dividend gives you access to more money each year, making your return that much better.

Larger premiums obviously mean more money at retirement and more growth as the internal rate of return is better.

As stated earlier, it's all relative. If you have a multi-million dollar operation the bigger numbers are possible. If you do not, then we have to go smaller. Simple as that. This comparison is not telling you $10,000 is not enough

Funding Comparison ROR

Year	Age	Premium	Cummulative Premium	Increase in Cash Value	Net Cash Value	Death Benefit	Rate of Return	Year Increase CV Equals Premium	Year CV Equals Accum Premium
1	46	10,000	10,000	6,910	6,910	492,638			
5	50	10,000	50,000	10,013	42,862	588,532			
15	60	10,000	150,000	14,730	167,778	791,585			
20	65	10,000	200,000	17,559	249,808	883,147	3.12	5	11

Year	Age	Premium	Cummulative Premium	Increase in Cash Value	Net Cash Value	Death Benefit	Rate of Return	Year Increase CV Equals Premium	Year CV Equals Accum Premium
1	46	20,000	20,000	12,393	12,393	912,452			
5	50	20,000	100,000	20,385	83,776	1,085,078			
15	60	20,000	300,000	29,668	336,463	1,450,782			
20	65	20,000	400,000	35,258	501,337	1,618,600	3.14	5	11

Year	Age	Premium	Cummulative Premium	Increase in Cash Value	Net Cash Value	Death Benefit	Rate of Return	Year Increase CV Equals Premium	Year CV Equals Accum Premium
1	46	40,000	40,000	27,702	27,702	1,250,164			
5	50	40,000	200,000	42,903	179,690	1,635,909			
15	60	40,000	600,000	62,750	713,081	2,451,072			
20	65	40,000	800,000	74,679	1,062,144	2,822,231	3.38	4	9

Year	Age	Premium	Cummulative Premium	Increase in Cash Value	Net Cash Value	Death Benefit	Rate of Return	Year Increase CV Equals Premium	Year CV Equals Accum Premium
1	46	70,000	70,000	46,581	46,581	1,997,405			
5	50	70,000	350,000	75,798	312,549	2,646,191			
15	60	70,000	1,050,000	110,411	1,252,619	4,020,004			
20	65	70,000	1,400,000	131,247	1,866,318	4,649,538	3.39	4	9

to start. It is a great place to start if you are just starting out yourself. What it is showing for those of you who are dealing with larger sums of money and have the availability to fund at a higher rate, is that you will benefit much faster should you invest larger sums of money. Focusing on getting the most for your money inside this concept does not happen by putting in less, but by putting in more.

If you are farming land and making a 3.12% rate of return and have the ability to make 3.39%, would you make some changes? I am guessing you would. Your operation is based on profit, not loss.

WHAT IS YOUR NEXT STEP

Next you implement your choices we talked about back on page 5.

Just as you don't go to the hardware store or the pet store for farming and ranching advice, you should not go to your local friend, neighbor or general life insurance agent to set this up if he/she has not heard of this concept and utilized it. I understand you like all those people and already have a relationship with them, but this is not about them and it's not about me—it's about YOU and YOUR financial future. You need to think about whether you want this done correctly or whether you want to take yet another chance with your money? Remember, farming is already a risk.

Your next step is to contact me, or the agent who gave you this book, to make an appointment and see if this will work for you. I can't imagine it won't, but one never knows until we visit.

Do Your Due Diligence

If you do not visit with me or the person who gave you this book, you better be asking the following questions of the person you decide to work with:

> Q. Do you know what whole life is?
>
> Q. Do you recommend whole life?
>
> Q. Do you know what a Paid Up Additions rider is and how it works?
>
> Q. Do you know what Modified Endowment Contract means?
>
> Q. Do you work with mutual companies who pay dividends?
>
> Q. What is the financial rating of your company?

If they answer the above questions and you hear any of the following comments or words, you better either run or think long and hard about what you are doing:

- "Yes in our "whole life" you can make a 8-12% rate of return." Remember, some agents call Universal, Variable and the mixes of the two, whole life. They have the terminology wrong – those are permanent life products.
- Surrender charge
- No guarantees
- Death benefit should increase over the years

- Illustration should show at least 60% of your premium available in cash value year one.
- If they do not work with mutual dividend paying companies.
- If their company rating is lower than an A-.
- "Whole life agents line their pockets with commissions."
- "Whole life is expensive, term is the better way to go and invest the difference."

What if you already have a policy?

1 Be sure that policy is a true whole life policy and not one of the other permanent products I talked about earlier on page 36 and above. The majority of the people I talk to tell me they have whole life, **only** to find out it is not whole life, it is one of the others. If you purchased your policy after 1980, I will bet it's a Universal or Variable product. The sure sign is if it shows you a "surrender value". Whole life does not have surrender values.

2 If you have a true whole life policy, good for you! You are HALF way there. Most likely your policy was set up for all the premium to go to death benefit and, yes you can use it, but you are not going to get anywhere fast. It is good but very inefficient and will not give you access to the

> amount of money you want or need. Should you cancel it? NO! These are good policies; what we may do is start another one for you that is set up correctly and built for financial growth as well as death benefit.

It may look easy to someone who does not understand, but it is not that simple. The process, coupled with the product, is the key and you want someone who is an expert. This is a time you need to be selfish. **For** far too long you have not had control, and it's time to take it back slowly.

Procrastinate no longer. **Remember,** waiting costs you money. If you are reading this, and it's fresh in your head, then contact me now to set up a time to visit. Again, my phone number is 701-751-3917, email is maryjo@fiscalbridge.com and the website is www.farmingwithoutthebank.com.

Summary

Farming has advanced in many ways, but the way you are financing that farm has not changed for centuries. In order to move forward you must change the way you think about money, just as you have changed the way you think about farming practices.

Nothing happens overnight. You must be able to look forward years to see the results of what you have built when you are using a whole life insurance policy.

The benefits are greater than the wait for huge results. Remember you are gaining:

- ✓ Immediate access to money
- ✓ No loan qualification process
- ✓ Payment flexibility during the tough times
- ✓ No more lost interest to banks
- ✓ Increased profits per acre
- ✓ Tax-free use of this money during retirement
- ✓ Leaving a family legacy
- ✓ Ability for farms to stay in the family

This concept is not an easy one to grasp on the first read, so read and re-read this book several times to get a handle on the information being relayed to you. I also encourage you to read *Becoming Your Own Banker* by R. Nelson Nash. After you have done this I encourage you to call and visit with me or another certified IBI practitioner. Agents all across the country will tell you they know what this is and they can help you, yet they have no concept of how to set up a policy for your benefit. Just as you wouldn't go to the local garden store for crop planting advice, you do not want to go to an agent who is unaware of the dynamics of this concept.

I thank you for reading and hope you found this book to be worthy of your time. Please do feel free to send me comments, as I would love to hear them and answer any questions you may have.

Contact me at maryjo@fiscalbridge.com, visit my website at ***www.farmingwithout thebank.com*** or give me a call at 701-751-3917. To send me "snail mail" please check the contact us link on the website for my mailing address.

Recommended Readings

Becoming Your Own Banker by R. Nelson Nash
Building Your Warehouse of Wealth by R. Nelson Nash
Financial Peace of Mind by Dwayne Burnell
Financial Independence in the 21st Century by Dwayne & Suzanne Burnell
How Privatized Banking Really Works by L. Carlos Lara and Robert P. Murphy
Great Wall Street Retirement Scam by Rick Bueter

About the Author

I, Mary Jo Irmen, was raised on a farm in western North Dakota where my parents raised purebred Charolais cattle for over 39 years and farmed. My dad was that young rancher, who bought his first cow when he was an eighth grader. Today my parents are farming with my brother and sister-in-law.

Struggles in farming and ranching are always present. I have heard stories of my great-grandparents and have watched my father and brother. Working with other farmers has only emphasized the issues I saw early on when I was introduced to the concept. Farmers are running large business operations. They are running multi-million dollar corporations and paying interest to someone else for their operating money.

I have yet to come across anyone who deals with the YEARLY operating expenses similar to what farmers deal with. It's a good year when those expenses can be paid back in full and a stressful year when they cannot.

These are the same people feeding our country, yet they are only offered the same finance option as 150 years ago.

It is not something I want to see continue; it's time for a major shift in thinking and some preparation to change that thinking.

In 2008 I lost a large sum of money in the stock market. All I remember was calling my advisor (who is now an agent with me). She advised me to keep putting in money and not pull any out. I, of course, did not listen. All I could think at the time was what kind of stupid advice is that? Who would keep putting money into a failing opportunity? So I stopped contributing and started looking for ways to make money in the market. I would read books on how to invest myself and hand them over to my advisor just for her to say no.

When this concept was presented to me I was just blown away. I thought, how in the world can this be legal and no one know about it. I went home excited and figured out I better educate myself before I decide to put some money in this and lose it to these crazy people. So I read every book I could get my hands on, decided I was going to get licensed and, through that process, really understood that this WAS in fact very much legal. Not only is it legal, it is one of the best kept secrets because it cuts out a middle man for whom the farming industry is very lucrative.

What you see today is my passion for helping those people with whom I grew up—farmers and ranchers. I have made it my mission to change the way the industry is financed and create more stable financial systems for those individuals. As they are working hard to provide me with so much, I, too, will continue working hard to supply them with more profits and less stress.

Disclaimer

The illustrations in this book are for educational purposes only. Illustrations do not represent any particular insurance company or have any kind of guarantees associated with them. They are intended for teaching purposes only.

Numbers are not guaranteed, as results will change based on age, gender, health, state, company and type of insurance, as well as other conditions. This book does not represent any insurance company in particular.

Illustrations in this book assume payment of dividends for all years shown. However, dividends are not guaranteed and may be declared annually by the insurance company's board of directors. Dividends fluctuate, and, therefore, can be higher or lower then illustrated.

All illustrations were run as NO MEC's (modified endowment contracts) allowing all loans against the cash value to be income tax free.

The Infinite Banking Concept® is a registered trademark of Infinite Banking Concepts, LLC. FiscalBridge, LLC is independent of and is not affiliated with, sponsored by, or endorsed by Infinite Banking Concepts, LLC.

CPSIA information can be obtained
at www.ICGtesting.com
Printed in the USA
FSOW02n0557110215
5119FS